Testimonia

I found "Handling the Remedy" to be a bold initiative. It provides an innovative and practical approach in identifying customers' and their requirements through involvement of each and every personnel within the organization; using a well-designed questioning process. In today's fast changing environment, customer needs keep changing, therefore regular alignment with the customers is a necessity so that their current needs are being addressed and the changing trends are identified. An insightful and interesting read.
Timir Ghosh

———⟨∰⟩———

In today's environment of decentralized support models, ever changing technologies, shrinking budgets and the need to do more with less, Handling the Remedy begins by placing the focus back where it belongs. With almost as many standards to adopt as there are types of industries to apply them to, everyone rushes into the latest trend or industry mandate without remembering the first key to any organization which is an effective alignment of the team under a common understanding. More than any other challenge, this will be the one that will prevent any good company from becoming great and David tackles it head on and accurately. I anticipate the balance of the books and the insight they will bring.
James Bennett

———⟨∰⟩———

David – Thank you for putting into words an easy to follow process for identifying alignment issues between staff and customers and how to address these issues. In my experience, this is a systemic issue in many companies that is realized by many, but hard to define and quantify. Your book describes the issues and provides an easy to follow identification process. I look forward to the next book in the series!
Bill Thornley

———⟨∰⟩———

Handling the Remedy is definitely a paradigm shift. I have often wondered if there was a better way to handle a solution that would encompass a wider scope. There have been many situations where the people that worked on solving problems were so isolated in what they are doing or belong to a silo of responsibility that it prevents them from seeing the big picture and seeing the perspective of the customer. Handling the Remedy has taken this into account and proposed a solution. It is an interesting, informative and instructional read. I definitely recommend Handling the Remedy for anyone looking for an overall solution to organizational problems.
Doan Ong

———— ⊂≡≡⊃ ————

This book is a must read for those wanting to align their businesses in the most optimal way. Handling the Remedy will cause you to think with greater insight about your business; and to come to real, tangible understanding of the relationship between individual departments, the company at large and most importantly, the customer. The first step is admitting you have a problem, and Handling the Remedy will help CEOs and business owners everywhere, to see with absolute clarity what and where the problems are, and begin the process of optimizing their business and their people.
Shaun Jansen

———— ⊂≡≡⊃ ————

Handling the Remedy touches on a very real issue encountered in IT Operations – differentiating between incidents and improvements. When an organization fails to separate these two activities, it's the enhancements that will take a backseat to incidents. Therefore, recommendations geared towards keeping an organization ahead of their competition will struggle to see the light of day. As David explains, establishing two distinct groups of resources allows both incident resolution and improvement opportunities to be implemented in parallel.
Mike Doyle

———— ⊂≡≡⊃ ————

HANDLING
THE REMEDY

A new paradigm for aligning
business and technology solutions

David G. Peterson

NOTICE

DEDICATION

To my wife, Elaine, for holding down the fort and looking after the kids while I embarked on this long journey. To Ray K., for helping me get started and for all the advice. To my wife, family, friends, and editors for their comments, suggestions, and proofreading.

A big thank you! Oh! There's another book!!

Contents

Introduction

Whether you're an executive or a leader, a company president or a manager, at some point you have spent most of your day working on a technological problem. Technology is the backbone of your organization; it is both your business advantage and your business roadblock.

But how do you convert your technology problems into solutions? How do you go from project failures to success? How do you go from following to leading? How do you get various best-practice solutions to work in harmony? The answer lies within a new paradigm, which I have termed "Handling the Remedy."

Handling the Remedy is the *corpus callosum* – it brings left and right brain thinking together to achieve solutions that are aligned to organizational objectives.

Handling the Remedy is also a holistic methodology by which you can improve the organizational management of solutions through the proper alignment of people, processes and best practices, while building neatly on solutions already in place.

How much time and energy have you used trying to implement best practices in your organization? There's an inherent problem with best practices – think about it. Best practices are comprised of hundreds, if not thousands, of organizations that did an aspect of a process better than others. They were able to do that one task better because their organizational environment supported it. Those tasks were then arranged for other organizations to use as the best solution.

Now, there's nothing wrong with that – or is there? Remember, best practices are formed by various industries, including agriculture, banking, cosmetics, defense, energy, food, health care, legal, and more. Does your organization involve all of these industries? That's likely not the case! So, instead of adapting your organization to follow the best, align the best to adapt to your organization.

Has your company institutionalized the means for coordinating all undertakings across departments and for aligning all endeavors with customer need? Does the IT department focus solely on the technology of the solution leaving business to decide if organizational objectives are being met? These are among the questions I have addressed in proposing the institutionalization of Handling the Remedy.

Additionally, Handling the Remedy is a means for hard-wiring the measures that prevent organizational fragmentation. It is the means by which a company knows before a project is initiated whether the proposed solution will integrate with and leverage existing technology solutions. Handling the Remedy is the concretization of a managerial philosophy that fully recognizes the need for technology solutions to be more responsive to evolving customer need; it answers the companies' need to maintain their identity and core principles while taking on the structure of adaptability.

The overriding goal of Handling the Remedy is to achieve the best possible organizational alignment with the customer in the most judicious adoption of business and technology solutions. So, though Handling the Remedy is many things, above all, it is a structural feature that vastly facilitates communication across organizational departments while ensuring the greatest possible critical appraisal of the information being communicated.

To summarize Handling the Remedy is:

- The institutionalization of a questioning attitude.

- A holistic, comprehensive, and organic approach to manage solutions company-wide.

- Designed as a repository of recursive intelligence for each solution as it affects the entire enterprise.

- Intelligence that is constantly revised according to a continuous stream of feedback that meets the highest standards of objectivity.

- Both a *conceptual* framework to ensure the optimization of extant and future solutions in terms of the overriding objectives of an enterprise and the *realization* of this framework in concrete, organizational structures.

- The systematic governance of a remedy or remedies in the enterprise. It is the optimal handling of remedies. Just as solutions vary based on the kind of enterprise at issue, the handling of these solutions should also be business-specific.

Specialization of labor is a reality of economic life, with the same kinds of tasks grouped for efficiency within an organization. One widespread managerial practice today involves creating distinct domains of activity. Once these domains are defined, you can better refine relevant management techniques for the greatest yield. However, there is a danger whenever management loses sight of the whole.

In the IT department, for example, tasks are often divided into subsections, or divisions, such as Incident Management, Problem Management and Change Management. At the outset, these domains co-existed, and their subsequent separation has widely been considered a wise move. Before the adoption of Information Technology Infrastructure Library (ITIL) processes, for instance, IT departments struggled to maintain control over their own operations, often with end users and customers feeling the brunt of it. This was due to inadequate incident management and the failure to distinguish incidents from problems, a state of tangled affairs made worse by the absence of procedures for apprehending and properly addressing change.

IT departments today are much more complex than they were even just a decade ago, and the challenge now is "what" to manage. Handling the Remedy is crucial in identifying the "what" and ensuring that processes, procedures and people are aligned to achieve management's organizational objectives.

As business evolves, managerial compartmentalization can become profound. As one gains a better view of individual departments, the organization as a whole tends to go out of focus. In fact, it is sometimes questionable that compartmentalization *does* provide a better view. This would seem to be the case given an analytic model, but what if the various elements that constitute an enterprise cannot be isolated – what if they can only exist as part of a whole?

To determine how things really stand, the next time you end up spending most of your day in a meeting regarding a technology problem due to an incident or project, ask the people present these three questions:

1. List all the products and/or services your company offers?

2. Which products and/or services do you support?

3. How does what you do affect the bottom-line?

When employees are not aware of what their company offers, it means the organization is not informing and updating them on new strategies, future endeavors or products and services that are no longer available. There is a relationship between how employees are kept informed and managerial presuppositions. Employees without practice in taking a holistic view will fall into an atomistic perspective, and they will take this perspective into all areas of the organization. To have informed employees, one must have leaders who communicate with a comprehensive understanding that the employees can then widely disseminate.

The employee who cannot answer one or more of the three questions above is not alone, by any means. Most people find satisfaction in their work when they feel they are contributing to a part of the larger whole, but it becomes difficult to clearly see this contribution given an analytic model – the model that created managerial domains. The employee, too, must appreciate how what he or she does every day affects the company, and that means that C-suite and call center operator alike would benefit by a more dialectic understanding of the workplace.

A business is not like a picture puzzle that can be taken apart into discrete pieces and then reassembled. A business is much more like an organism. It is the nexus of constitutive relationships, and whether employees grasp this concept makes all the difference in the world. Handling the Remedy is a methodology that permits a given solution to receive the attention it requires without the organizational culture breaking down into silos.

ROADMAP

Handling the Remedy is comprised of three segments. The *first* segment is designed to determine the alignment between employees and an organization. The *second* segment is the framework, which coordinates the alignment of people, processes and best practices. The *third* segment looks at various best practices and how to align them to your organization.

This book focuses on the first segment - Alignment.

To achieve alignment you must first identify your three main customers:

1. The organization

2. The department

3. The division (the department within the department)

Figure 1 shows you your three customers and how they should be aligned to achieve success. It also shows how misaligned you could be.

The next three chapters comprise of exercises, containing seven questions each, designed to measure knowledge and understanding of these customers and to create a baseline.

These exercises are best done anonymously. Alternatively, at your next technology crisis meeting, ask the questions to everyone present and see how well they do collectively.

There are various ways to extrapolate and interpolate the data that you derive from these exercises; some creative ways are provided toward the end of this book.

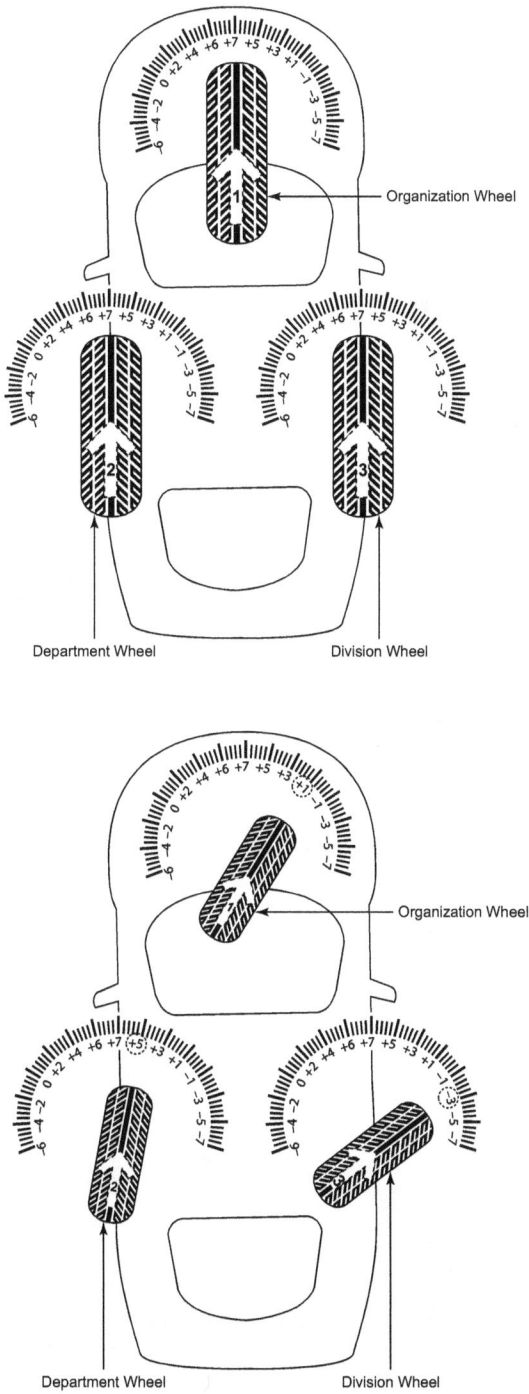

Figure 1

Q&A PART 1 - THE ORGANIZATION

No one should underestimate the power of the right question. You need to consider this now because the rationale for Handling the Remedy can best be understood by way of questioning.

When companies grow rigid and unresponsive, they lose the ability to adapt, and become entrenched in old habits of practice; there is resistance to change. Intelligent change can only begin by virtue of the right question. The right question opens up a "space" for what is absolutely requisite to innovation—decisive moments of uncertainty. The sclerotic company cannot tolerate uncertainty. The company that intelligently adapts to fluid circumstances always raises questions.

An indispensable presupposition of every question is a confession – namely, "I don't know." Many people and organizations would rather avoid such a confession because uncertainty provokes anxiety. But the uncertainty that arises here is an inescapable prelude to creativity. To do anything differently, you must be open to the humility of this confession and its accompanying anxiety—and then, ask the right question. As you move through the question-and-answer exercises below, you will perhaps face some hard realities about your workplace, but by facing them, you will see the urgent need for Handling the Remedy.

Let's get started. Go through the following questions and answer them as well as you can, spending no more than thirty seconds on each. If you are unable to answer within this time, skip the question and move on to the next. The questions will require that you consider the solutions that your enterprise makes available. If more than one solution is relevant to a question, list those additional solutions.

1. What solution (product or service) does your company provide?
2. What need does the solution address?
3. How does the solution benefit the customer?
4. How can your company improve the current solution?
5. When was the last improvement made to the solution?
6. What is the ideal, or best solution?
7. What prevents your organization from implementing the ideal, or best solution?

If you were unable to answer a few of the questions, you are not alone. However, the simple fact that you are now reflecting on the questions puts you far ahead of the pack.

If you do not directly interact with customers, some of the questions may have been challenging. As companies expand, so do the departments in which employees work. Before long, departments become business entities unto themselves, and it is easy for employees to forget what their own organization offers its customers.

Such employees are not aligned with the solutions in place. This may mean that the employees who provide support or implementation services within their departments do not contribute to the overall goals of the enterprise. Their departments have floated free from the context that originally gave them purpose, and the employee of a given department no longer sees his or her place in the whole. One of the reasons to adopt Handling the Remedy at your company is to prevent this narrowing perspective that inhibits and threatens growth.

The questions you answered above were designed to get you thinking about a wider perspective, one that would render your work more meaningful. If your company is losing its competitive edge, for example, your answer to these questions can be an enormous aid in identifying the likely culprits. Without such self-examination, there can only be repetition or imitation.

Given the accelerating rate of change across all business environments and the demands of vastly intensified competitive forces, there is a temptation among some to follow fads, aping the examples of other companies in expectation of similar returns. (Think of what one fast food chain is undertaking in the reproduction of its ambiance that hitherto has been peculiar to a premium caffeinated beverage chain.) Before falling into repetition—designing "new" procedures that remained moored in old assumptions—or imitation, first pause and raise the kinds of questions that broaden your perspective.

Now, imagine that you posed these same questions to the total five employees of a certain company. Naturally, the questions would be answered easily. A company that is properly integrated in this way can grow extremely quickly. It is both remarkable and informative how swiftly such "simple" questions can become difficult with company expansion.

Look at each question and consider what elements the answers should include:

Q1. What solution does your company provide?

A.   Your answer should include the names of all the products the organization sells and/or all the services it offers. Perhaps an employee who cannot provide a complete list of names or services is only personally involved with a few of them. However, there is nothing benign about the limited involvement that leads to insufficient knowledge of the company. Any solution such an employee implements is at risk of being problematic because some products and services are unknown. An integrated implementation can only take place within what is known and understood. Such an employee cannot consistently work toward the overall objectives of the enterprise.

For example, if you tell your employee, "Let's go north," I hope he or she asks, "Which north—magnetic or true north?" If some employees think you're referring to magnetic north and others think you mean true north, you have a problem, as the difference between the two is significant. Although the goal (going north) appears to be the same, the two results will send you to two entirely different places.

This is an important distinction that touches upon organizational communication.

Q2. What need does the solution address?

A. This is important. If employees understand what need each of your business's solutions or products addresses, they are aligned in their thinking and able to communicate effectively. When they fully appreciate the needs that the enterprise meets, they can work within their various departments and coordinate with other departments in the company without proving divisive and counterproductive to growth. They can design, implement or support solutions that are consistent with the organization's central mission – its self-identity. A corporation is no more exempt from the injunction, "To thine own self be true," than a human being. Corporate integrity looks rather like the human variety. It is attained when all of the parts perform their functions in service of the whole, according to the highest good.

Q3. How does the solution benefit the customer?

A. To answer this question, the employee must put himself in the customer's shoes. By being so situated, the product or service itself appears in another light. It is then easier to determine whether improvements need to be made, whether the item is sufficient, or whether the product or service provides an incontestable value. A space must be created for critical reflection, and the space required is that of a question. To raise the right questions is to court the uncertainty by which it is possible to think otherwise, to think as one is perhaps not in the habit of thinking. Creativity, innovation, and originality are impossible without the uncertainty that attends a departure from old ways of thinking and behaving. The company that is afraid of uncertainty is on automatic pilot. Too many companies have lost their way by losing track of how things look from the customer's perspective. This perspective must be regained.

Q4. How can your company improve the current solution?

A. Once employees have put themselves in the customers' shoes, improvements will follow with a certain logical inevitability. For example, a customer reaches the end of his or her online transaction and receives the recommendation to print it for his or her records. The option to save a PDF file of the transaction might be a benefit (a convenience), especially if the customer is not able to print at that time or simply prefers the electronic alternative. The customer could be given the option of receiving the receipt as an e-mail to file in his or her mailbox. Once the employee understands the customer's perspective, maximization of utility is only a matter of attentiveness. No inspired insight is required.

Q5. When was the last improvement made to the solution?

A. The answer to this question will show you how your organization handles a given solution. Improvements illustrate the capacity for growth and reveal that customer needs are understood. Businesses need to update their solutions constantly to continue to appeal to their customers and meet their evolving needs.

Q6. What is the ideal, or best solution?

A. Imagine that a company offers a new product or service that is the best of its kind. The company is just as thrilled as the customers who enjoy the new product or service. A danger, however, can occur when the company's self-satisfaction renders it complacent and unable to appreciate how the product or service can change, just as the consumer does. What had been ideal at the product's inception may no longer serve its original purpose, and the solution now may need to be updated or replaced entirely. To foresee customer need adequately, a company must be committed to its consumers enough to embrace them for an entire life cycle—and that implies a certain flexibility.

Q7. What prevents your organization from implementing the ideal, or best solution?

A.   Management is not always delighted to hear such a question, but an enlightened manager understands that roadblocks are either needlessly disabling or signposts to success. Limited budgets and resources are common constraints, but they leave open the issue of proper allocation, a determination that can only be made holistically. Nagel, a famous philosopher, has written of "the view from nowhere." In a world of human beings, there is no view from nowhere. But, as paradoxical as it sounds, there is a "view from everywhere." A healthy organization must be able to see itself in the round. It cannot permit its view to be compartmentalized. Long contemplation of the customer—in the form of the most advanced possible gathering of information and solicitation of involvement—is an indispensable means for a robust organizational self-understanding.

A successful organization addresses the changing needs of its customers. Looking at itself from a customer's perspective an organization can view itself as a whole. Everything depends on how the organization handles the remedy. Does it rush changes into production to get ahead of the competition, or does it hold out to ensure that it releases a quality solution?

Has it built critical reflection, the space for raising questions, into its very structure? (In significant ways, the strength of an enterprise can be measured in how well it can countenance questions without suffering paralysis.)

Handling the Remedy ensures that items are produced at the right time and in terms of the kind of critical reflection that represents a leveraging of the intelligence the enterprise already has on hand but, without Handling the Remedy, is probably frittering away. A questioning attitude is the first step to stemming this tide of waste.

The questions enumerated above provide insight into the level of awareness that employees have achieved with respect to their companies. You now need to consider things from the department's perspective.

Q&A Part 2 - The Department

Have the questions posed thus far made you uncomfortable? If so, that is a good thing, as it means you are learning and recognizing your company's strengths and weaknesses. Handling the Remedy involves raising questions. It recognizes that raising the right questions is as much an art as answering them thoughtfully. Everyone with a curious three-year-old knows that questions can sometimes make people uncomfortable. But if you are still reading and have endured the creative discomfort of these questions, you can take satisfaction in knowing that you have taken the first crucial steps down a path that will lead to greater customer responsiveness and permit you to manage your company's products and services with maximum alignment.

Now, begin with a new perspective. Approach the questions from a departmental point of view. Consider the following questions from the perspective of the department to which you belong. (Let's say your organization's marketing department includes an Internet marketing division. If you work in the Internet marketing division, answer these questions from the perspective of the marketing department.) Also, if your department provides more than one solution, give separate answers for each.

Spend no more than thirty seconds answering each question. Thoughtful answers to these questions will enable you to form a realistic impression of your department, and this is a key moment in the comprehensive analysis presented in this book.

1. What solution does your department provide?
2. What need does the solution address?

3. Who receives your solution?

4. How do they benefit from your solution?

5. Whom do they contact to request improvements to the solution?

6. Whom do they contact for problems with the solution?

7. How can the current solution be improved?

Q1. What solution does your department provide?

A. Your answer should include a list of all the services or products your department provides or supports. Your list may be incomplete. If it is, this may mean that the solution you implement does not take into account all that the department provides. The solutions offered by a department are normally interrelated; understanding the nature of this mutual dependence is indispensable. The right solution is always contextual.

Q2. What need does the solution address?

A. Think about whom the solution benefits. Only in this way can you determine what is truly essential in your department. However, what is essential in your department can only be defined in terms of its relationship to the customer, and the customer, with his or her changing needs, is a constant reflection upon the organization as a whole. Thinking carefully about departmental relationships as they affect the customer is the best possible means for arranging solutions in a hierarchy, or assigning a higher or lower priority to each solution.

Q3. Who receives your solution?

A. This could be a customer or another department within your organization. Your department may provide services that other departments use to support the organization's customers. Services that you can map back to customers are critical to the organization. Addressing this question helps the department to allocate resources, focus on what matters, and understand how a solution is used and by whom.

Q4. How do they benefit from your solution?

A. This puts you in the customer's (or department's) shoes. Knowing how they benefit will ensure that you focus on what matters to them. Things then follow a certain logical inevitability. If certain solutions become outdated or unimportant, a natural transformation takes place. The solution has evolved to be more customer-aligned, leaner, and more efficient.

Q5. Whom do they contact to request improvements to a solution?

A. When people don't know whom to contact for improvements, it means two things: an avenue for suggestions for solution improvement is closed down, and the customers' changing needs are no longer being tracked. When a company encourages customer feedback, an extremely valuable asset, it has the chance to craft its image and deepen loyalty. Respect goes to companies that are transparently responsive. For years, a major hamburger chain has associated its corporate image with the opportunity to "have it your way." When customer suggestions are implemented, it means that they are being heard and that they are valued. How many times have you visited a company's website only to search in vain for a customer service phone number? It is clear that some companies do not want feedback.

It is possible to consider a product or service from the customer's vantage point, but only the customer can walk the necessary miles in those shoes. Some aspects of the customer's experience simply cannot be anticipated. As such, one must hear it directly from the customer. Of course, this means one must also be a truly expert listener.

Q6. Whom do they contact for problems with the solution?

A.   Determine how many different places people must go to report problems. If they have to go to more than one, that is too many. In this respect, companies should follow the "keep it simple, stupid" (KISS) principle. If your department doesn't have a general place where problems can be reported, then the conduit to resolve them is closed.

Q7. How can the current solution be improved?

A.   If nothing comes to mind, this does not mean there is no room for improvement. It is generally possible to envision improvements, even if they cannot be directly actualized, given other constraints. The improvement thus envisioned can become a valuable objective, but its implementation ultimately will depend on conditions becoming right for it.

These questions constitute an exercise intended to give the individual employee insight into his or her own knowledge of the department as well as the relationship of this department to the rest of the company and the customer.

Q&A Part 3 - The Division

Part 2, inquired into the solutions offered by the department as a whole. In this part, you will seek answers from your division's perspective. (In the previous example, the marketing department had an Internet marketing division. Now, answer the questions presented from the viewpoint of the Internet marketing division.) If the division provides more than one solution, give separate answers for each. Spend no more than thirty seconds on each question.

1. What solution does your division provide?
2. Who uses your solution?
3. How do they benefit from your solution?
4. Whom do they contact about problems with the solution?
5. When was the last improvement made to the solution?
6. Whom do they contact to suggest improvements to the solution?
7. What are the obstacles your division faces regularly?

Q1. What solution does your division provide?

A. The answer should include the names of all the products and/or services your division provides or supports. Do not leave any services out. If you are unable to identify all of them, it could mean that employees in your division are working in silos.

Q2. Who uses your solution?

A. When you answer this question, you will find out whether and how your division actually supports other divisions and/or customers. It may amaze you to discover how much work your division does for other divisions and how little benefit actually extends to the customer. The more a division directly provides services to customers, the more it influences the organization's catalogue of solutions. Divisions that exist purely to support the organization need to demonstrate that they know and recognize their ultimate customers.

Q3. How do they benefit from your solution?

A. Answering this question will reveal the degree of connection your division maintains with the customer. If the customer benefits from the solution your division provides, then the importance and value of your processes needs to be realized.

Q4. Whom do they contact about problems with the solution?

A. Your answer to this question should identify the person responsible for providing or fixing the solution. If there is no such person, that could be because fixing the problem is not an option currently. However, the reasons for the absence of such an option should be continually reviewed because the need to entertain novel options evolves rapidly. Whatever the case, you should consider the message your company delivers when a contact person is absent or unavailable.

Q5. When was the last improvement made to the solution?

A. If you don't know when the last improvement was made, either none was made, or it was not communicated. Whenever you make changes, everyone needs to be informed – this shows that you have listened to concerns, made the changes, and value feedback. These improvements and their corresponding messages are vital because they take place at the lowest level of the organization.

Q6.   Whom do they contact to suggest improvements to the solution?

A.   Almost all the time, the person whom your customers contact to suggest improvements is the same person they contact when they encounter problems. I hope you can see what is wrong here. A problem occurs when something doesn't work as designed, and an improvement is an enhancement to an existing product or service. A successful organization splits these responsibilities. In this way, both improvements and resolutions to problems can be implemented at pace with customer requests.

Q7.   What are the obstacles your division faces regularly?

A.   Examine what hampers your productivity and slows down your division's momentum. You can use "Handling the Remedy - Framework" to address these issues.

If you are functioning well in your division, these questions should have been easy to answer. If, however, you found them difficult, you are probably more isolated than you should be.

COFFEE BREAK

After completing the questions in Parts 1, 2 and 3, take a break to process the experience. Then, take out a pen and start scoring yourself using figure 2. Because each chapter was limited to seven questions, you can easily calculate the results over a cup of coffee.

Assign each answer a value of +1 or -1.

Part 1: For questions 1 through 5, give yourself one point for every complete answer, and deduct one point for every incomplete or missing answer. For questions 6 and 7, give yourself one point if you have an answer (whether correct or not), and deduct one point if you left it blank.

Parts 2 and 3: For questions 1 through 6, give yourself one point for every complete answer, and deduct one point for every incomplete or missing answer. For question 7, give yourself one point if you have an answer (whether correct or not), and deduct one point if you left it blank.

The highest possible score for each part is +7, and the lowest is -7. The highest total or combined score from all three parts is +21, and the lowest is -21.

Worksheets with automated score calculation are available for download at www.handlingtheremedy.com/resources

Question	Part 1 (Organization)	Part 2 (Department)	Part 3 (Division)
1			
2			
3			
4			
5			
6			
7			
Total			
Combined Score			

Figure 2

Interpretation Part 1
The Organization

Part 1 illustrates how your perceptions align with the organization as a whole. The answers will help you to determine whether you understand what the organization is trying to accomplish and, just as important, whether you understand your customers. You will normally see a profound difference between the answers provided by customer-facing and non-customer-facing staff. Employees who directly interact with customers are more likely to be aware of the organization's products and services. As employee interaction with customers decreases, so does employee awareness of products and services. Employees with higher scores in Part 1 are also more likely to score higher in the next two parts because they know whom to approach regarding improvements and problems. They understand the company, the customer profile, and what needs to be done.

These scores also provide insight into how well internal communications work and whether the organization is well integrated and primed for the future. Generally, employees who work in a high-performing, healthy organization score higher. The leaders of such organizations communicate frequently and effectively with their employees, informing and interacting with them about new or additional projects or processes. Such an enterprise is distinctively characterized by a culture of openness.

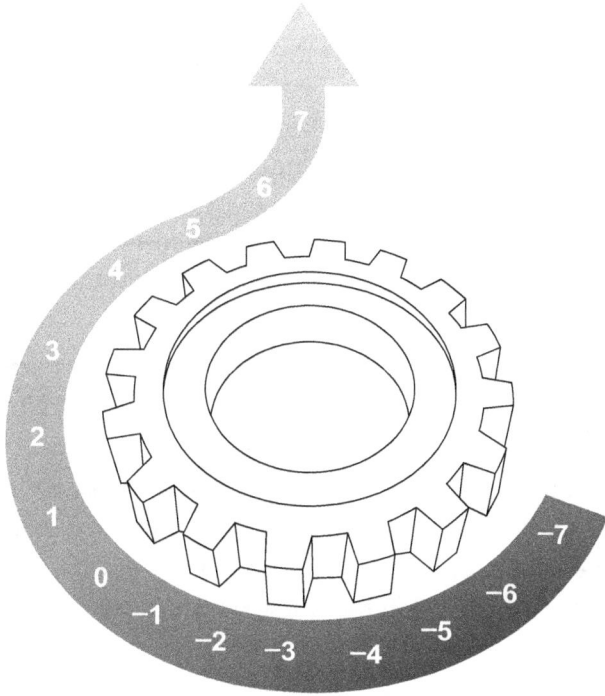

"*A positive score indicates that your answers are in accord with the strategic goals and overall mission of your organization.*"

INTERPRETATION PART 2
THE DEPARTMENT

Part 2 tests how a department's size affects its ability to operate successfully. When departments get too big and have multiple internal divisions, employees are often unaware of all the products and services the organization provides. Departments in which solutions are effectively aligned with the organization tend to have employees who score higher on these questions. Employees working in such departments know each other well, have regular meetings, and even schedule social events.

A poor score often indicates that the department is too big and hasn't rooted itself within the greater organization. In this case, information does not flow across the lines of specialization. Imagine that you enter a hospital and want to find the radiology department. When you ask a nurse where it is, he or she replies, "I'm not sure. Check with information." If you happen to catch the nurse on his or her first day, that's an understandable reply, but if he or she has been employed for a while, it is a symptom of a problem.

The smaller the department, the more closely people interact with each other. A division within a department can be all unto itself, with the employees living out their working lives in an isolated manner. A poor score could indicate the existence of silos. It could also indicate employees do not know how their jobs contribute to the overall effort of the organization.

7
6
5
4
3
2
1
0
−1
−2
−3
−4
−5
−6
−7

"A positive score indicates a cohesive departmental structure."

INTERPRETATION PART 3
THE DIVISION

Part 3 highlights the inner workings of a division. Is everyone doing their part to keep the "engine" of the division running, or is the engine slowing down and likely to come to a grinding halt?

A good score speaks to integration and a productive division. A poor score suggests the need for change. Divisions with employees who score low on these tests can affect other divisions. Needless to say, high-scoring employees can have a far-reaching positive influence on the organization.

After answering all of these questions, you should be able to name features of the workplace that would benefit from a change.

"A positive score indicates that your division is aligned with the needs of its customers."

The responses to the questions in all three chapters provide a snapshot of how your organization is doing; it's like a company health check-up. The correct answers to all the questions, especially those that bring scores down, should be included in training programs. You will need to revisit these at a later time, when the needs of your customers change and the organization's goals evolve to ensure continued alignment.

You can also repeat these exercises to check the effectiveness of your actions to achieve alignment.

Paint by Numbers

Now that you have some numbers, let's have some fun. This section is as interesting as it is eye-opening—just as when an art teacher gives you paints and paper and you realize that mixing primary colors produces other colors.

Use the following illustrations to visualize your division within the organization and assess it's performance. These are just examples; you may need to modify them according to your particular circumstances. Every detail counts, and it's important to paint the right picture. For example, Diagram A is best used with the scores of an individual or a single division or department, but Diagram B and onward will require the scores of various divisions in your department and/or organization to be used effectively.

Diagram A

It's crucial to align a vehicle's wheels for it to go in a straight line. The Q&A exercises earlier helped you take readings for each "wheel" of your organization.

In Diagram A, the Organization Wheel refers to your results in Part 1 of the Q&A exercise. The Department Wheel corresponds to your results in Part 2, and the Division Wheel corresponds to those in Part 3.

Go back to the scoring table, take the total from each part, and circle the appropriate number for the respective wheel.

You will see whether the wheels are lined up to steer you straight, or whether they will take you off your path altogether. Where are the wheels in your organization taking you? Even one degree of misalignment makes a difference. If you think of each wheel as a ship and you ask the captain of each ship to adjust his or her bearings by one degree, he or she will tell you that each ship will end up in other waters entirely. Over time, one degree makes a huge difference. Similarly, where would your organization's solutions end up?

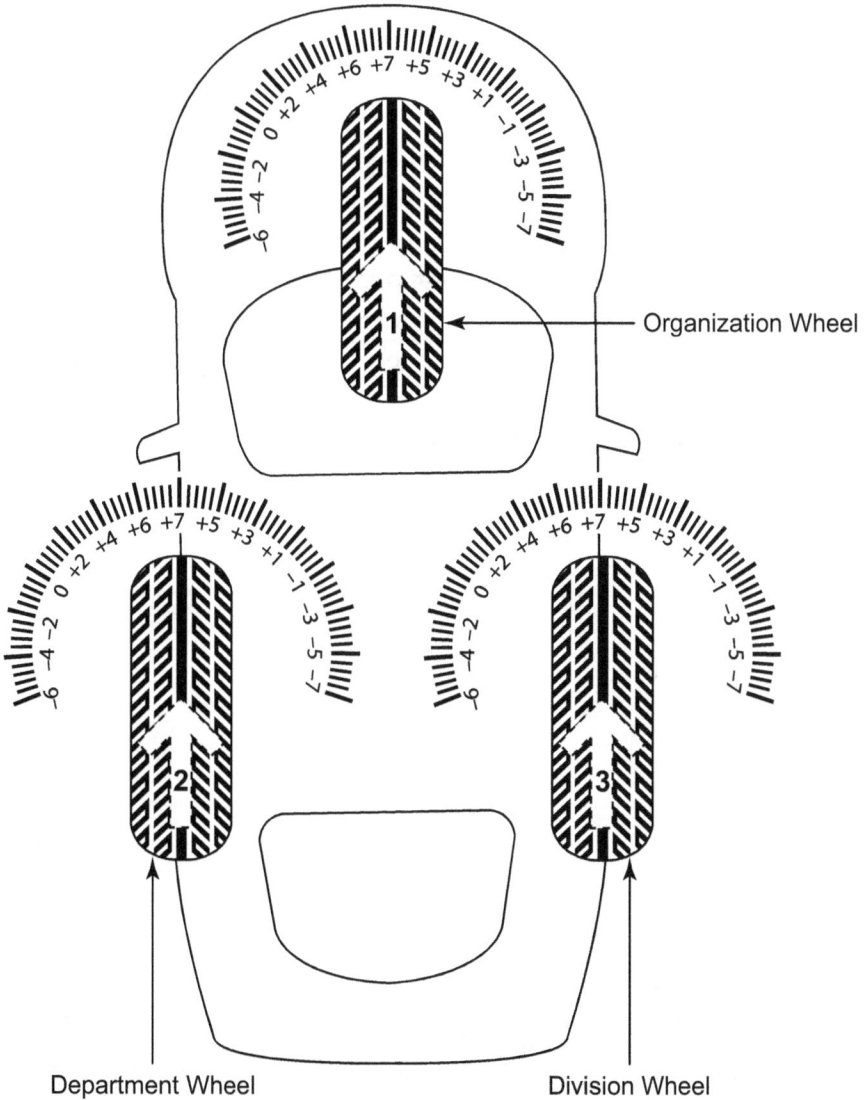

Diagram A

Diagram B

Diagram B represents a department that consists of three interacting divisions. (If the divisions do not interact with each other, use Diagram C.) Fill in the combined score for divisions A, B and C. To get the interdepartmental score, add the scores of each respective division. To get the department score, add up the scores for all the divisions.

What do you think of the score? Is a particular division increasing or decreasing your department score? If there are three divisions in your department, the maximum score you could achieve is +63 or -63 (because each division can only have a maximum score of +21 or -21). How was your interdepartmental score? As diagrammed, the maximum interdepartmental score is +42 or -42.

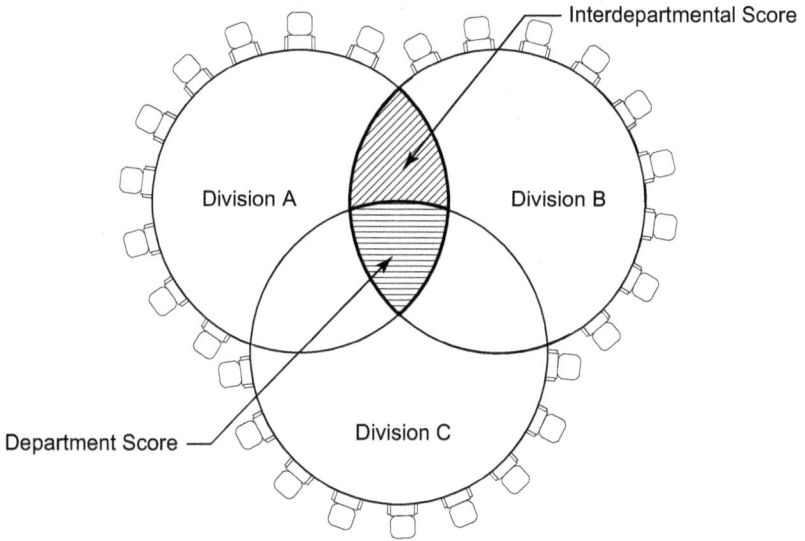

Diagram B

Diagram C

For departments with divisions that work independently, Diagram C works better. It represents a department that consists of four isolated divisions. Fill in the combined scores for A, B, C and D, and add them together to determine the departmental score.

This diagram also illustrates how the divisions working independently score within a departmental context. Does one division bring the score down for the entire department?

If each division has a positive score similar to that of every other division, this represents a department that is healthy and well balanced.

There is another interesting way to use this diagram. You can use it to score each of your organization's isolated departments. Now, notice how the organization performs with isolated departments.

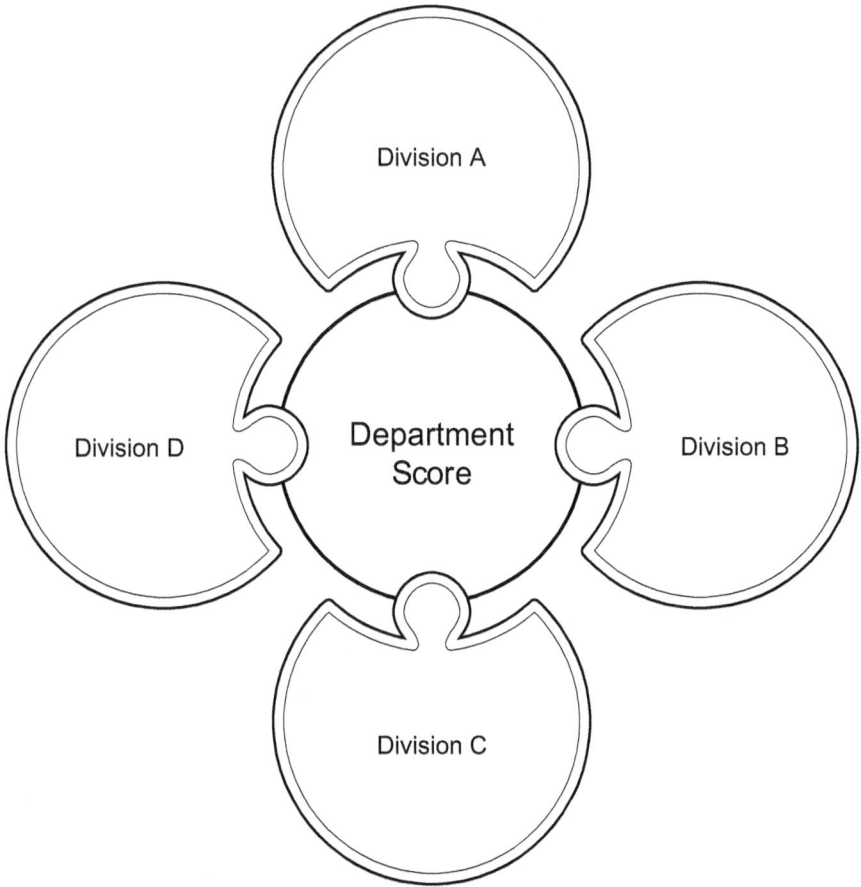

Diagram C

Diagram D

Not all departments conform perfectly to the previous diagrams. Diagram D might illustrate your particular department. Fill in the combined score for each division. Feel free to draw the intersecting circles that most accurately represent your department. To produce an interdepartmental score add the scores of the interacting divisions. To get the department score, add the scores for all divisions.

This diagram illustrates a complex department comprised of several divisions, and makes it easier to identify which division(s) bring the score down for the entire department.

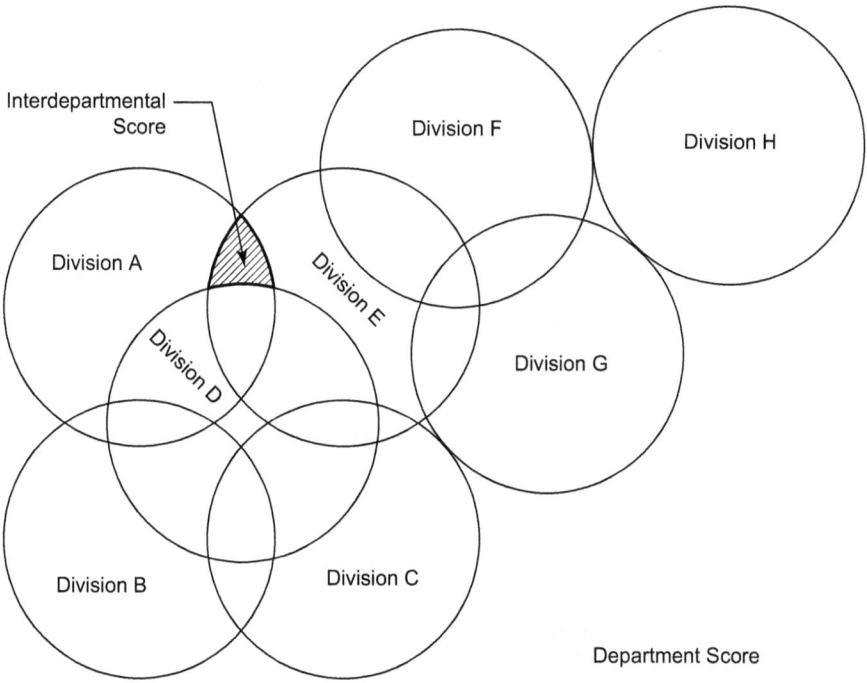

Diagram D

Diagram E

Diagram E is reminiscent of the "pass the parcel" game. The role played by each division in the life cycle of a solution is placed in chronological order, from start to finish.

Fill in the combined score for divisions A, B, C, D and E. To calculate the passing score, add the scores of the two respective divisions.

To obtain the final score, add all the passing scores. Again, the maximum score for each division is +21 or -21, so the maximum passing score is +42 or -42. For the illustrated example, the maximum final score is +168 or -168.

Just as scoring a goal in hockey depends on the players successfully passing the puck to each other, successful solutions require effective handling and the proper movement from one division to another. If the "puck gets dropped," this process helps identify where it happened.

Sometimes the process requires the same divisions' participation at different stages of the process. Draw Diagram E as it pertains to your organization.

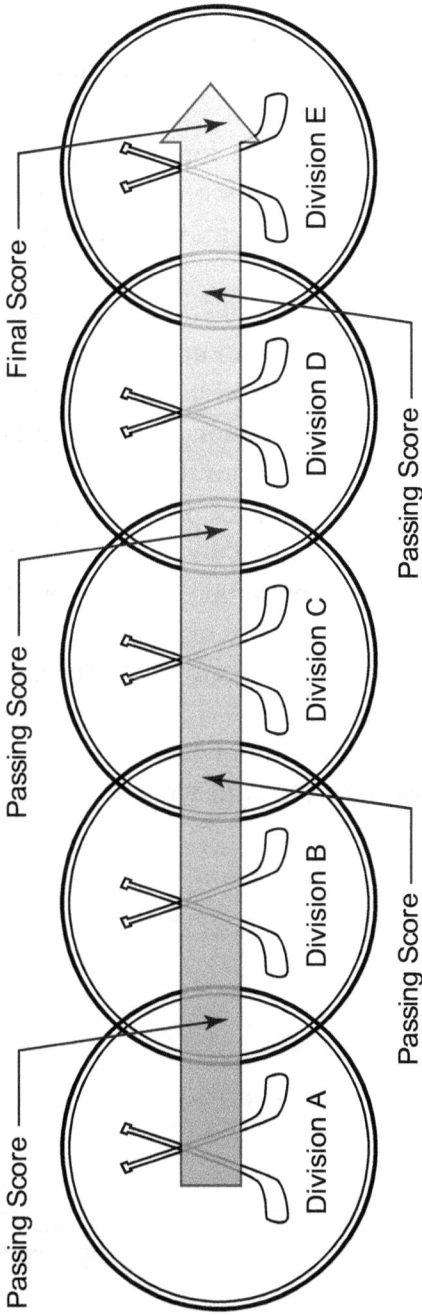

Diagram E

A chain is only as strong as its weakest link or, in this case, an organization is only as strong as its weakest division.

Diagram F

Diagram F displays all the divisions that contribute to your solution. It shows how each division handles the remedy and how that affects your division's solution.

Fill in the combined score for contributing divisions A, B, C and D. To get the output score, add up the scores of the divisions. In this example, the maximum output score is +84 or -84.

The diagram should be applied to every output stream for each division to identify areas for improvement. If you subtract the output score from your division's combined score, you will see how much harder your division works to achieve the overall objectives. For example, say your division makes pencils, and four divisions are responsible for providing materials, testing, resources and support. Say that the materials division is a high scorer and provides the materials on time, but the resources and testing divisions don't have high scores, and your work is delayed because of it. This diagram reveals trouble spots in the production of solutions.

Use this diagram to identify which divisions make positive or negative contributions to your work.

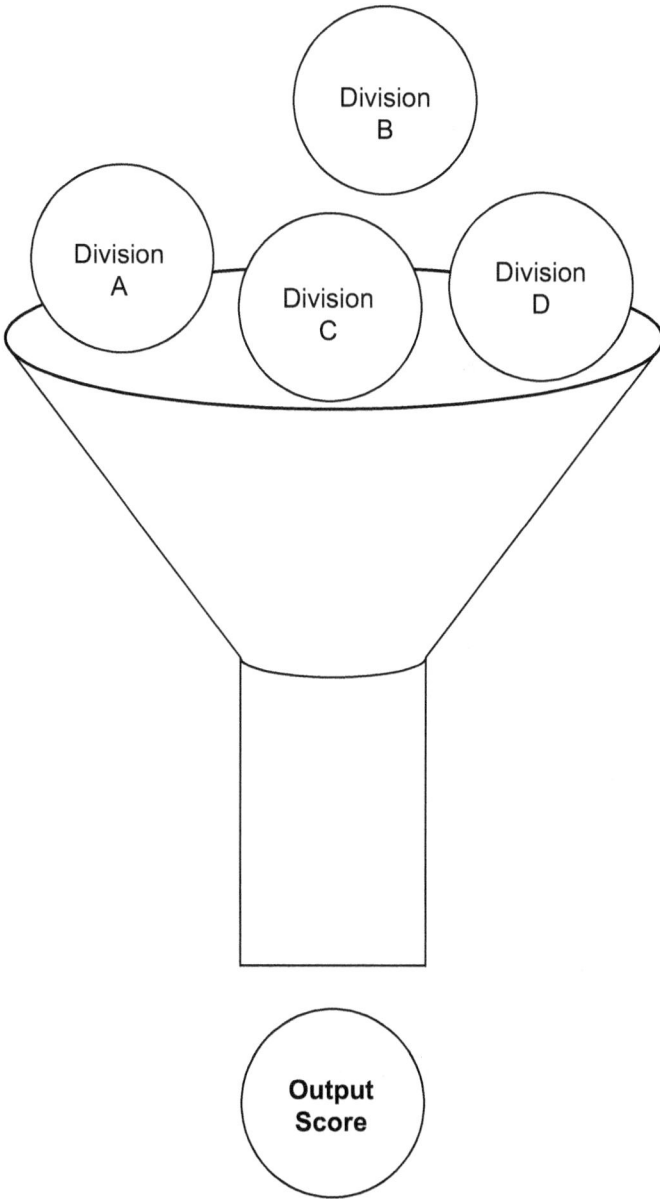

Diagram F

Diagram G

Diagram G shows how your organization as a whole handles the remedy by using the combined scores from each division. You need to answer all the questions in Parts 1-3 for all the divisions in your organization and use the combined scores generated there for this diagram.

Fill in the combined score for divisions A, B and C. To get an interdepartmental score, add the score of each relevant division. To get the department score, add the scores of every respective division. To determine the organization score, add all the department scores. In this diagram, the maximum department score is +63 or -63, and the maximum organization score is +315 or -315.

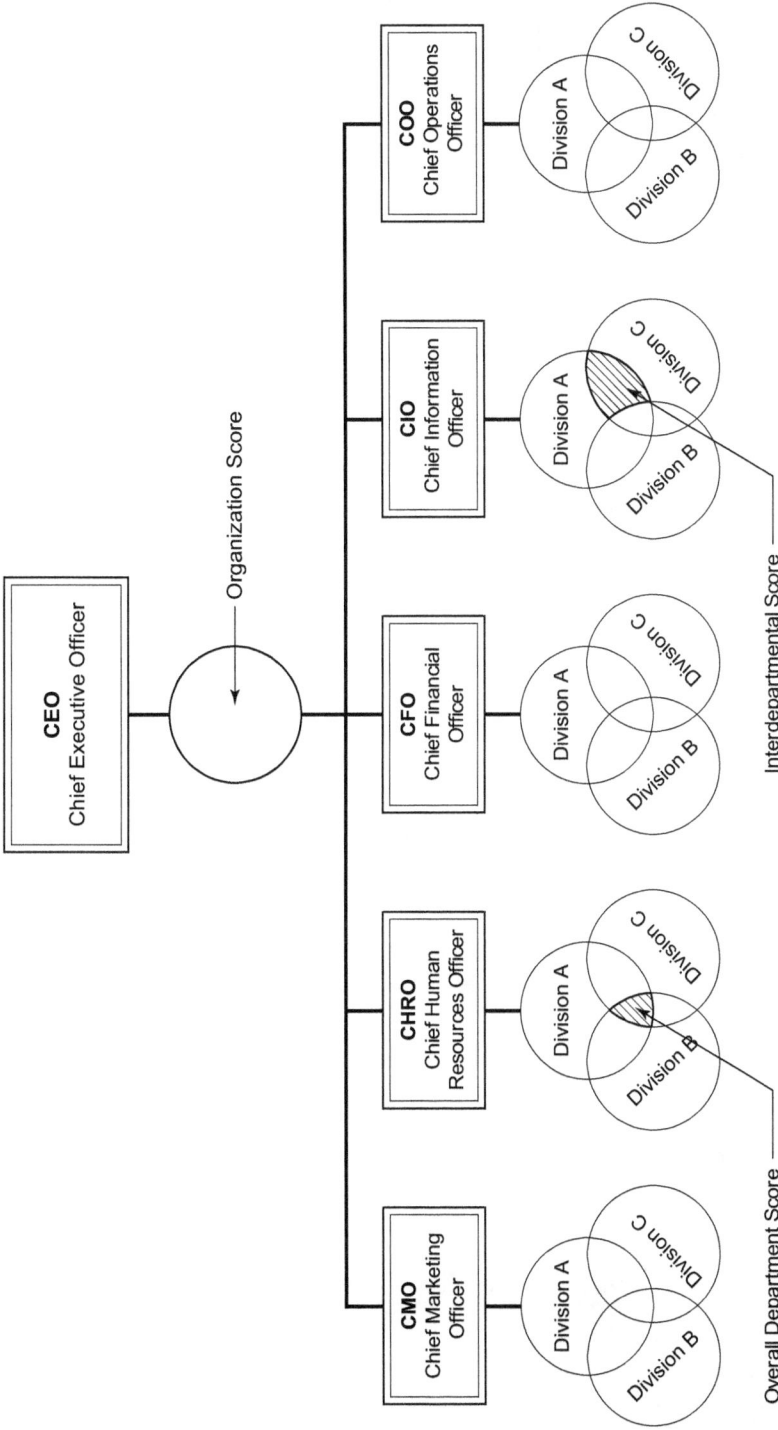

Diagram G

Diagrams A through G provide a quantitative, and visual method for gaining insight into how well divisions and departments operate throughout the organization. The diagrams disclose a network of relationships that might otherwise go overlooked and that could, in all likelihood, be better managed.

Summary

Handling the Remedy is a new paradigm that brings alignment to an organization through the three P's; people, process(es), and practice(s).

This book, the first in a series, examines the "people" aspect within an organization. It is designed to motivate employees to think, question and align with the goals of the division, department, and organization. With that clarity, everything follows a natural progression. Process and practice can then be correctly aligned. Through this methodology, resistance to change (the biggest barrier to an organization's success) is defeated through logical outcomes.

It's difficult to disagree with what is logical, and logic that comes about through self-realization is even harder to discount. When each individual goes through the three exercises, he or she will begin that realization process. At the end of that process, employees can plot their results on the three-wheeled car (Diagram A). That diagram will help them identify the areas in which they need alignment, and by repeating the exercises periodically, they can check if they have achieved alignment.

I invite you to share your experience and thoughts about this book via my website **www.handlingtheremedy.com** or e-mail me at davidpeterson@handlingtheremedy.com

Appendix: Case Study Example

Company XYZ's IT department consisted of three hundred FTEs (Full Time Equivalents) who worked together to ensure the company's technology services were always online. One afternoon inside the company's operations center, a display showing the status of online services and resources, such as networks, servers, mainframes, and firewalls flashed green, yellow, and then red.

A crowd of people streamed into the operations center. The CIO asked the head of operations what was happening. The question trickled down until everyone in the room was asking the same thing. Surprise turned to irritation as the failing services did not come back online and there were no answers.

Incident managers got involved, and experts from each division huddled into a room. The operations staff confirmed that some servers had restarted, communication links had gone down, firewall logs showed unusual entries, and primary servers had failed over to secondary servers. Users started to call the service desk, complaining that they couldn't log in to the necessary services.

At this time, the CIO and head of operations asked for an update that was not "geek speak." They were shocked to hear "we don't know."

The CIO asked for a status update of what and who was affected. Server administrators quickly listed several servers that were offline. The head of operations asked which services and lines of business were affected by those servers. The administrators listed a few services, but didn't know the lines of businesses, as there was no record of who used those servers.

Members from the Incident, Problem, and Change teams started reviewing their logs to find a reason for the outage.

Does this scenario sound familiar? To prevent such occurrences management decided to adopt the alignment process of Handling the Remedy. Fifteen individuals were selected to participate:

- Five members from the Incident Team, including three incident managers and two employees who monitored various systems and services.

- Five members from the Problem Team—three problem managers and two other employees who had participated in resolving a recent problem.

- Five members from the Change Team—three change managers and two Change Advisory Board (CAB) members who were responsible for approving changes.

All fifteen members were asked to complete Q&A Parts 1, 2 and 3. (The values for each member in this example were selected randomly, and the results averaged.)

These are the Incident Team's results:

Question	Part 1 (Organization)	Part 2 (Department)	Part 3 (Division)
1	1	-1	1
2	-1	-1	-1
3	-1	1	1
4	-1	-1	1
5	1	1	-1
6	1	-1	-1
7	-1	-1	-1
Total	-1	-3	-1
		Total	-5

These are the Problem Team's results:

Question	Part 1 (Organization)	Part 2 (Department)	Part 3 (Division)
1	-1	1	-1
2	1	1	-1
3	1	-1	1
4	-1	1	-1
5	1	1	-1
6	-1	-1	1
7	1	1	1
Total	1	3	-1
		Total	3

These are the Change Team's results:

Question	Part 1 (Organization)	Part 2 (Department)	Part 3 (Division)
1	-1	1	1
2	-1	1	1
3	1	1	-1
4	-1	1	-1
5	1	1	-1
6	1	-1	-1
7	1	1	-1
Total	1	5	-3
		Total	3

Diagram 1

The readings shown in Diagram 1 come from the Change team's Q&A exercise—so much for going in a straight line!

This diagram represents the alignment to the organization, the department and division. You can see that the Change team is going in a different direction from its parent department and the organization.

Using this diagram and the question-and-answer scoring, you can identify the areas that need to be aligned to help the change team proceed in the same direction as the rest of the organization. Teams can repeat this exercise later to check progress.

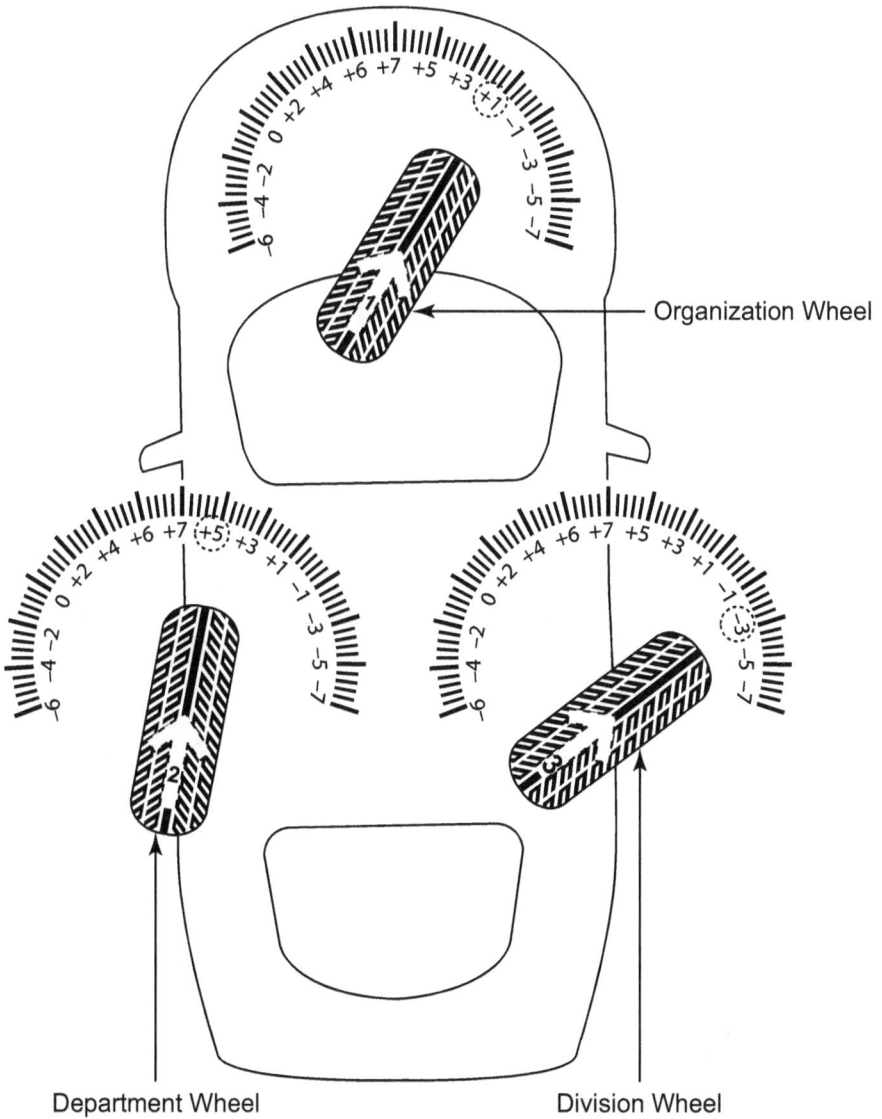

Organization Wheel

Department Wheel

Division Wheel

Diagram 1

Picture three cars—one each for the Incident, Problem, and Change teams—and put them side by side. Based on the earlier scores, if the cars took to the road, they would crash into each other!

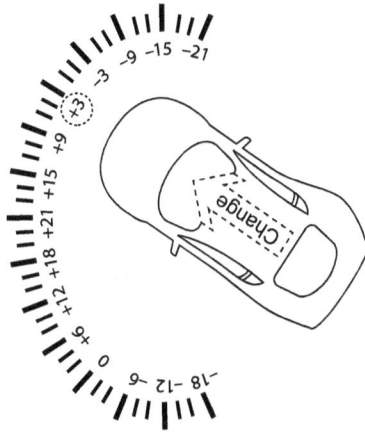

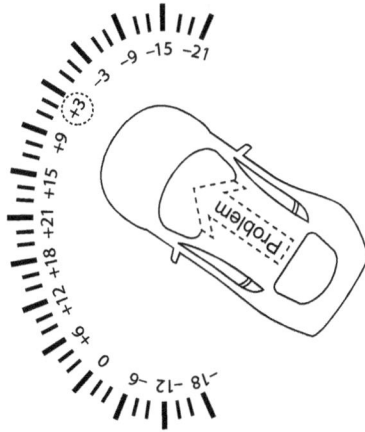

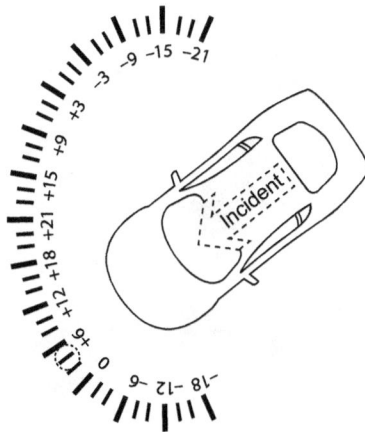

Diagram 2

By plugging in the scores, you can see how each team stacks up and how they stack up against each other. In the example below, the Change team and Problem team have a combined score of +6, but when either of those teams work with the Incident team, their combined score drops to -2. The possible scores between the two teams range from -42 to +42.

The department score is positive (+1), with a possible range of -63 to +63. From the scores, it is clear that the positive influence of the problem and change teams is masking the issues faced by the incident team. You can clearly see that the incident team should be the first focus of the organization.

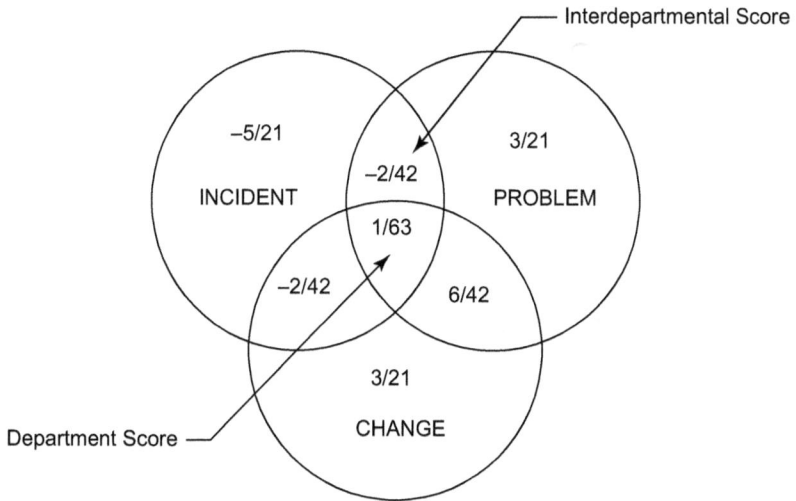

Diagram 2

Conclusion

From the case study described earlier, you can see the need for Handling the Remedy. In organizations, a person or team is tasked with trying to resolve issues. Normally, issues within the control of the division or department are fixed, while those that require interaction with other divisions and/or departments do not get the required "traction" and are buried under newer problems.

This is where Handling the Remedy framework comes in. It compiles data from various divisions and departments, tracks and analyzes activities that fall out of their jurisdiction and provides results that are aligned, coordinated, and communicated across the enterprise.

Handling the Remedy is the arbitrator for cross-divisional processes, ensuring that they are fixed across all divisions and not partially resolved by the division with the originating problem.

Handling the Remedy methodology supports executives by allowing them to lead an aligned productive organization that ultimately improves the bottom line.

About the Author

David G. Peterson has extensive international experience managing projects and operations for large financial institutions. He has worked in North America, Europe, Middle East and Asia skillfully managing business and technical requirements, core systems enhancement and support, merger and acquisition integrations, business process reengineering, off-shoring and outsourcing.

Visit his website **www.handlingtheremedy.com** to find articles, his blog, videos, testimonials, and more.

www.handlingtheremedy.com

NOTES

NOTES - Q&A PART 1

1. What solution (product or service) does your company provide?

2. What need does the solution address?

3. How does the solution benefit the customer?

Notes - Q&A Part 1

4. How can your company improve the current solution?

5. When was the last improvement made to the solution?

6. What is the ideal, or best solution?

7. What prevents your organization from implementing the ideal, or best solution?

Notes - Q&A Part 2

1. What solution does your department provide?

2. What need does the solution address?

3. Who receives your solution?

NOTES - Q&A PART 2

4. How do they benefit from your solution?

5. Whom do they contact to request improvements to the solution?

6. Whom do they contact for problems with the solution?

Notes - Q&A Part 2

7. How can the current solution be improved?

Notes - Q&A Part 3

1. What solution does your division provide?

2. Who uses your solution?

3. How do they benefit from your solution?

Notes - Q&A Part 3

4. Whom do they contact about problems with the solution?

5. When was the last improvement made to the solution?

6. Whom do they contact to suggest improvements to the solution?

Notes - Q&A Part 3

7. What are the obstacles your division faces regularly?

www.ingramcontent.com/pod-product-compliance
Lightning Source LLC
Chambersburg PA
CBHW071110210326
41519CB00020B/6254